Daniel Graves

A Treatise on Sericulture

Compiled from Some of the Best Authors Extant...

Daniel Graves

A Treatise on Sericulture
Compiled from Some of the Best Authors Extant...

ISBN/EAN: 9783337280680

Printed in Europe, USA, Canada, Australia, Japan

Cover: Foto ©berggeist007 / pixelio.de

More available books at **www.hansebooks.com**

SERICULTURE

COMPILED FROM SOME OF THE BEST AUTHORS EXTANT;
BEING A GOOD TEXT BOOK FOR THE NOVICE
AND FIRST BEGINNER.

BY DANIEL GRAVES,

PROVO, UTAH COUNTY.

PRINTED AT STAR BOOK AND JOB OFFICE,

SALT LAKE CITY.

1880.

PREFACE.

—

N the compilation of the following Tratise on Seri-culture, I have been careful to avoid all extraneous matter, making it so plain that a child or illiterate person can understand its contents, and be enabled to manipulate the silk-worm in all its varied stages. I have selected from the most reliable authorities on the subject, and have added some notes coming under the notice of persons en-gaged in sericulture in this Territory. The treatise contains: 1st—a synopsis of the early and present history of silk cul-ture; 2nd—the variety of trees; 3rd—various kinds of silk-worms; 4th—their diseases, with causes and remedies; 5th—the concoonery, hatching, feeding, etc.; preservation of eggs and cocoons.

DANIEL GRAVES.

INDEX.

TREATISE ON SERICULTURE.

SYNOPSIS OF THE HISTORY OF SERICULTURE.

IT is stated in the oldest books of the Sanscript that the culture of the silk-worm commenced first in China in the reign of Houng-to, third emperor of China, and known in history as the Emperor of the Earth, B. C. 2700, being cotemporary with Joseph, the son of Jacob, while sojourning in Egypt. Houng-to conceived the idea of bringing into use the insects, which were indigeneous and fed on the leaves of the wild mulberry tree in that region, and requested his empress, Siling-chi, to prepare a place and collect them together, and feed them. It appears she did this, and not only fed them, but spun the silk and made it into garments; the Empress Silingchi˙ is defined as the "Goddess of Silk-worms," for the great boon so given to her country, and at the present time, as has been done from time immemorial, the Empress, with the ladies of the Court and ministers' wives, go in the springtime to the fields east of the city, purify and worship the goddess; they are not allowed to wear costly apparel, or work in embroidery, during the time for the cultivation of the silk-worm. At the present time it is one of the greatest employments of that people, who export millions of pounds of raw silk annually, as well as millions of dollars, worth of eggs. From China it extended to India, Japan, Arabia and other nations of Asia, Europe, and lastly to America. At the present day the silk-worm is becoming one of the greatest sources of wealth to those contries in which it is cultivated.

It was not known to the ancients from what silk was obtained until Aristotle conjectured that it was unwound from the pupa of a caterpillar.

In the reign of the Emperor Aurelian of Rome, A. D. 280, an attire of silk was considered too great a luxury for even an empress to wear, being worth its weight in gold. The mode of producing and manufacturing silk was not known in Europe until the sixth century, when two monks of the order of St. Basil arrived at the court of the Emperor Justina at Constantinople, on their return from a mission to China, A. D. 556, who brought with them in their casques, seeds of the mulberry and the silk-worm.

Its culture soon commenced in Greece, and the Venetians were soon able to supply the whole west of Europe. It was soon introduced into Spain, Portugal and Sicily, and in 1540 it extended to Piedmont and through Italy, where it soon ranked with the best of Asiatic origin. In France, near Aragon, in 1340, mulberry trees were being planted; in 1499, it was introduced in Alan, but was not thoroughly established until 1603, when Henry of Navarre encouraged it. The Huguenots were largely engaged in the silk business during the reign of Louis XIV, but owing to the Edict of Nantes being revoked in 1685, nearly 400,000 were driven into exile, and as many more perished and were slain. This almost annihilated the silk manufacture in France. In Lyons, 18,000 looms were reduced to 4,000, and in Tours from 10,000 to 12,000 looms were destroyed; its mills were reduced from 800 to 70. Weavers could not be found for them. Over 100,000 Huguenot refugees fled to England.

M. Mavet asserts, in his history of the silk trade, that the first mulberry tree was brought to France during the crusades of Guipage of St. Aubon, and planted three leagues from Montmeliart. This tree was said to be living in 1810, when M. de la Tours de pay le Chaux, caused a wall to be built around it, forbidding its leaves to be gathered. The cuttings and descendants of this tree now cover the soil of France and produced to the state in 1810, more than 100,000,000 lbs. of raw silk, and more than 400,000,000 francs in this industry only, and the amount has greatly increased since that time.—*Count DeHazzi.*

The consumption of silk in France in 1873 was quoted at 1,600,000 lbs., and the production in 1812 was 987,000 lbs. of raw silk, and a like quantity was imported. Dr. Lardner, Count de Hazzi, states as follows: "This single* branch industry in 1836 was estimated at 40,000,000 florins; a tenth of which was obtained from the production of its raw material, and the remainder from its manufacture. The kingdom, from recent statistics, derives from its silk products 23,000,000 francs per annum, and 84,000,000 francs from its fabrication; whereby the capital brought into circulation amounts to 108,560,000 francs, and now it has become one of the greatest industries. In 1865 the value of silk goods produced was estimated at 106,000,000 francs, of which 26,500,000 francs' worth of raw material was imported. The home consumption was 35,000,000 lbs.; export 71,000,000 lbs.; silklooms employed 225,000, giving food to half a million persons. One-third of these products was brought to the United States; this country paid France for her silk goods, $9,900,000. Finizio, the celebrated manufacturer of Naples, makes and sends to the New York market 3,000 lbs. of sewing silk per week.—*Silk Cul. vol xvi, p.* 133.

"Turkey," says Dr. Landey, "supplies England with a considerable quantity of silk; the imports from that country averages over 333,000 lbs. annually. It also supplies 3,000 lbs. of sewing silk annually. In Turkey its productions are confined to large cities and towns, and in their neighborhood the mulberry trees are raised by the farmer, who carries on a lucrative business by gathering the leaves and taking them to market every morning during the season of feeding the worms. The principle place in Asia Minor for silk is Broosa, about ninety miles south of Constantinople, which, in favorable years produces from 7,000 to 8,000 bales of raw silk. At the beginning of the season every family clears out all the rooms but one, in their houses, in which they live.

Mr. Russell, in his interesting volume on Ancient and Modern Egypt, says: "In the valley of Tumulate, the ancient land of Goshen, is established a colony of about five hundred Syrians, for the purpose of cultivating the mulberry tree and raising silk worms, in the beautiful province of Fayoum, forty miles south of Cairo, on the west side of the Nile."

The average amount of silk annually exported from Italy is computed to exceed $17,500,000.—*Count Dondolo on Silk Culture, vol. i, p.* 15.

About the year 1605, James I encouraged the raising of the mulberry tree and silk worm, but the climate was found to be unhealthful for the insect, and he turned his attention to the Colonies. In 1622 the Virginian Company was urged to promote the cultivation of the mulberry tree and the raising of the silk-worm, and instructions were given to the Earl of Northampton to urge the cultivation of silk in preference to that of tobacco. He wrote to the governors to compel the planting of mulberry trees, which met with the concurrence of the Assembly in 1623, and every land owner was compelled to plant at least ten trees for every one hundred acres of land he possessed, or pay a fine of ten pounds of tobacco. In 1646, an act was passed, describing silk culture as the most profitable industry in the country. The King, having worn the Virginian silk, told Sir William Berkely, then governor, that it was not inferior to that of other countries.

In Georgia, in 1732, a piece of ground was given by the government for a mulberry plantation and nursery, and a native of Piedmont sent to instruct the people in this branch of industry. In 1735, eight pounds of raw silk was sent from Savannah to England, where it was woven and sent to the Queen. In 1749, to encourage the growth of colonial silk, it was exported free of duty, and an Italian gentleman was engaged to instruct the Colonies in the Italian method of management. In 1758 the weight of silk cocoons received at the filature was 1,052 pounds; next year 7,540 pounds; the following year, 10,000 pounds, and at that time the raw silk from Georgia, sold in London from two to three shillings per pound more than that from other countries. In 1768 the business was greatly increased, when a serious disaster befel it. The storehouse took fire and consumed a large quantity of raw silk and about 8,000 pounds of cocoons. The last parcel of silk brought for sale at Savannah in 1790, upwards of 200 pounds, was purchased for exportation at from eighteen to twenty-six shillings per pound.

The silk culture in South Carolina commenced about the time it did in Georgia, in 1732, and became a fashionable employment for the ladies of Carolina, as it should be with others. The celebrated Sir Thomas Lombe said that the quality and excellency was equal to any produced in Italy. In 1768, Dr. Franklin recommended to the American Philosophical Society of Philadelphia, a filature of raw silk at Philadelphia, which was built by private subscriptions in 1770. The following year 2,300 pounds of raw silk was brought to reel, and in 1828, by order of Congress, a work was published which says that in 1771, silk culture commenced in Pennsylvania and New York, and continued with spirit for several years.

In 1770, Mrs. Susanna Wright at Columbia, Lancaster County, made a piece of matua, sixty yards in length from her own cocoons, which was afterwards worn as a court dress by the Queen of Great Britain. Grace Ischen, about the same time, made a considerable quantity of silk stuffs, a piece of which was presented by Governor Dickinson to the celebrated Catherine Maccerly. Many ladies, prior to the Revolution wore silk dresses of their own fabrication. At a large meeting of silk growers in the hall of the Franklin Institute, January 7, 1839, Dr. Mease presented several specimens of silk woven and dyed about the time before mentioned, which was highly creditable, and the venerable speaker facetiously remarked, though not so splendid as that now manufactured, nothwith-standing it would make a very fine show in a country church.

In 1760, Mr. N. Aspinall planted the white mulberry tree and introduced the eggs of silk worms in the town of Mansfield, Conn., and New Haven, and Dr. Eyra Styles obtained a bounty from the legislature to encourage the raising of mulberry trees and raw silk.

In 1789, 200 pounds of raw silk was produced; in 1793, 365 pounds. In 1810 the sewing and raw silk of New England, Windham and Tollard were valued by the United States marshal at $2,850, exclusive of the amount of domestic fabrics, and double that amount in 1823. So popular indeed had the silk products become at this period that they were readily taken and paid there as a circulating medium. In 1839 the New England States were more or less engaged in the culture of silk and its manufacture, which was encouraged by the bounty given by the legislature to encourage the silk enterprise.

From 1780 to 1823-4, the amount of silk made in the United States was not very great. It had been more of a domestic manufacture in several districts; many families made there from five to fifty pounds annually, and an enthusiastic worker brought her produce from eighty to one hundred pounds. This domestic produce was more common in Connecticut than elsewhere, but was quite common in New Jersey and in a few other states.

During the War of 1812-13, Samuel Chidsey, of Cayuga, New York, sold sewing silk annually to the amount of $600, from silk raised by his own family.

In 1875 the United States paid $28,540,369 for raw silk. At the present time the silk industry of the United States is making great strides, and bids fair to compete with the other silk growing and manufacturing countries. In many of the old tobacco growing states they are plowing up their fields and planting mulberry trees.

PART SECOND.

CHAPTER I.

On the class and order of trees, with the Genus Morus, commonly called the mulberry tree, of which there are five species,. namely: The Alba or White Tartarian; Nigra, or Black; Rubus, or Red; Tinctoria, or Fusticwood.

Species 1. Alba, white; China; leaves slightly cordate, equal at the base, ovate or lobed, unequally serrated, smooth, consisting of ten sub-varieties.

Species 2. Tartarica, Tartarian, Tartary; leaves slightly cordate, equal at the base, ovate, or lobed, equally serrated, smooth.

Species 3. Nigra, black, or common; Italy; leaves cordate, ovate, or lobed, unequally toothed, scabrous.

Species 4. Rubia, red; North American; leaves cordate, ovate, accuminate or three lobed, equally serrated, scabrous, soft beneath, fine spikes, cylindrical.

Species 5. Tinctoria, Fusticwood; West Indies; leaves oblong, unequal at the base, spines auxilary, solitary.

Morus Tinctoria is a tall branching tree, with a fine head, smooth leaves, and oval shaped solitary spikes. The whole plant abounds in a slightly glutinous milk, of a sulphurous color. The timber is yellow, and a good deal used in dying, for which it is imported under the name of fusticwood.

Morus Multicaulis, which has been highly recommended by Mr. Hugg, of California, and others, is entirely condemned by the United States Silk Association. (See History Silk Industry, published 1876, chapter 6, page 38. The Morus Multicaulis Mania.) It was easily raised from cuttings. Gideon B. Smith, of Baltimore, is said to have owned the first tree in the United States, which was planted in 1826; but Dr. Felix Pascalis, of New York, was the first to make known to the public the remarkable growth and supposed excellence of the

tree, and so opened the Pandora's box, from which so many
evils followed. The excitement grew slowly until 1839, when
it progressed so rapidly as to cause ruin to its cultivators. The
young trees or cuttings which were sold in 1834-5 for $3 or $5
per hundred, were worth $25, $50, $100, $200, and even $500
per hundred.

In the spring of 1839, when Mr. Whitmarsh and Dr.
Stebbin, of Northampton, were rejoicing over the purchase of
a dozen Multicaulis cuttings, not more than two feet long and
of the thickness of a pipe stem, for $25, exclaimed the Doctor
in enthusiasm, "they are worth $60." A nurseryman, who in
1835 had sold small quantities to nurserymen in several of the
Massachusetts cities, whilst at work, determined to make a
push for a speculation; he took and dressed himself up and
went to all those he had sold to formerly, and others that had
any of said trees. Arriving at Newport they offered fifty
cents each for all he had. The nurseryman thought a moment, and then raising up his head, said: "I do not think I
want to sell what few I have;" and he went from place to
place. "I came back," said Mr. ———, "and you could not
have bought in any of the towns I visited for a dollar apiece,
although a week before they would have been satisfied with
25cts. each for them." After having imported from France
and multiplied his cuttings, and from the success he had
made, so enormous were his sales, that in the winter of 1838-9
he sent an agent to France with $80,000 in hand, with orders
to purchase one million or more trees, to be delivered in the
summer and fall; but the crisis had come before their arrival
and the bubble had bursted; so great was the panic that no
purchaser could be found. Thus was the great mania and
downfall of the raising of silkworms in the states, the ruin of
many, for hundreds of acres were planted with trees which was
of little account. Supposing the Morus Multicaulis was what
has been claimed for it by some, it would not suit this north-
ern clime, being too tender to stand our winters.

The Morus Alba is the hardiest of all the varieties, and is

said to contain more silk than any of the other classes; but there are many sub-varieties of the first named, for by sowing seed, they, like many other things, do not produce their like. As in other fruits, the best way is to raise by layers, as cuttings do not succeed well.

CHAPTER II.—PREPARING THE GROUND FOR PLANTING.

1.—Plow or dig deep, making furrows and filling them up with good rotten manure before planting trees or cuttings, and when planted press the soil tightly to and around the roots. Compost leaf, mould, and rotten stable manure are suitable.

2.—In sowing the seed make a rich bed facing the south, with good composts, similar to the former, only mixing sand to make it light; keep the soil moist; the time of sowing is from the 1st of April to the 1st of May, but in some seasons until the 1st of June. It is computed, says a writer in the first volume of a silk manual, that one ounce of seed properly sown will produce 5,000 young trees. Judge Compstock says from one pound of seed may be realized 100,000 trees. He also says there are about 322,700 seeds in a pound; this therefore allows one seed out of every three to vegetate. Count De Hazzi says: "From 9,600 to 10,000 seeds weigh one ounce of our Bavarian weight." Judge Compstock says: "Before sowing, the seed should be steeped in water about blood warm for from twenty-four to thirty-six hours," according to an actual trial made on the 9th of February, 1839. Let the plowing or digging be done the previous winter and leave it rough; should the weather be dry, water every other evening; water with liquid manure or soap suds once a week. Sow in drills, stirring the ground between the drills prior to watering with

the liquid manure or soap suds. The first winter protect the
young plants by mulching them with litter or straw, first lay-
ing across some large willows so as to keep the litter or straw
from pressing too heavy upon them, as soon, or before there
comes a black frost.

3.—Setting out a mulberry orchard. Says Mr. Goodrich,
president af the Hartford Silk Association: "I advise to set
the rows eight feet apart, and in the rows two feet apart, which
will take 2,700 trees to an acre, and when they are too thick
to take out every other one so as to plow between the rows,"
and he also says, "plant potatoes between the rows until the
roots and tops begin to spread too far, the ground being well
manured."

In India and Persia the dwarf orchards are not allowed
to rise above eight feet, being so pruned that a man standing
upon the ground may be able to reach the top. Begin to
prune the young trees the first year after having set them out,
to three eyes, and so on each year, the young twigs being used
to feed the worms. Let no leaves grow below two feet from
the ground. Mr. Bonafout, the celebrated writer on silk culture, the disciple of Count Dondolo, and the director of the
Royal Gardens of Paris, and also Dr. Tinally, D. C. S., Univer-
sity of Paris, says: "This method is generally adopted in
Italy and the plan of hedgerow planting is done with less ex-
pense and more profit than standard planting."

CHAPTER II.—PLANTING.

The planting of Morus Alba in the hedge form will be
found the most advantageous; the same quantity of land will
thus produce at least 80 per cent. more leaves than from
standard trees, and the labor for gathering full one-half less.
Count Dondolo says: "The mulberry tree should only be

stripped once a year, and that crop should be gathered so as to allow time for the leaves to shoot before the cold weather sets in, or the tree will shortly die. The planting out of a mulberry orchard is one thing and the raising of silk another; for, as in France, Italy, Brooza, Turkey or Persia, the orchards are farmed out and the leaves brought to the cities, (where the worms are raised by women and children,) and sold every morning. But here in Utah, where nearly every family have a small piece of ground by their dwellings, trees may be set by their fences, so as to let the children pick the leaves and feed the worms. This is what may be termed cottage culture, whereby every family of four or five children, in about six weeks can earn more than all the clothes they wear would cost. Let them be planted from the fence two feet and apart three feet.

Several years since a farmer in the vicinity of Mansfield, Connecticut, purchased a farm on which was planted twelve mulberry trees, of full growth. Knowing nothing of the business of making silk he supposed them to be of no more use than the ordinary value of forest trees for fuel. A neighbor, however, called upon him and agreed to pay him twelve dollars annually for the privilege of picking the leaves. The farmer, to his astonishment, found that the twelve trees were as good to him as $200 at six per cent. interest per annum.

A writer to the *Silk Culturist*, inquires whether a farmer cannot plant trees, and let them out to poor families, to make silk on shares. The answer given is as follows: "There are but few farms but what might be fourfolded, and by adopting this course would give an opportunity to the industrious poor, not only to provide for the present wants of their families, but to lay up something in store for the day of adversity.

Note.—In gathering the leaves, be sure to leave leaves at the top ends of every twig to be preserved to draw the sap, and preserve the life and vigor of the tree. And in gathering

the leaves be sure not to strip by drawing the hand downwards, for by so doing you may destroy the germ or bud at the
root of the leaf, but be careful to pass the hand upwards.

PART THIRD.

CHAPTER I.—GENUS, SPECIES AND VARIETIES OF THE SILK WORM.

The silk-worm, or Bombyx Mori, is one of the various
families of caterpillars that pass through several transforma-
tions into their final state, the moth or butterfly ; order, lepid-
optera, being four flaked wing insects and are silk producing
The silk secretion of the caterpillar of the Bombyx Mori
is said to be provided with glands, by which the juices
of the mulberry leaf are discussed and secreted so as to supply
the different organs without any admixture of other ingredi-
ents. Without this, the silk would not be of the quality and
texture in which it is found. According to Randohr, these
secretors consist of two transparent membranes, between which
flows a yellow, limpid jelly. The longer the secretors, the
greater is the quantity of silk expended by the insect in the
construction of its cocoon. When the larva of the Bombyx
Mori attains its full size, it ceases to eat and instinctively prepares and encloses itself in three coverings: 1— with floss ;
2—with silk ; and 3—with gum ; with which last it lines the
inside of the cocoon, except at one end, which it only partially
closes ; and at that end where it will have, from the position
of its body, neither inconvenience nor obstruction when the
period arrives for it to make its egress.

The cocoon being constructed, disengages itself from its fourth skin and enters its chrysalis state; after throwing off the above skin, it makes a new one, which hardens into a leathery hide. In ten or twelve days the chrysalis swells, bursts, and the moth struggles out of its leathern envelope into the chamber of the cocoon, and makes its way out as a moth with wings to float in air.—*Spectacle de la Nature; Count Dondolo* and others.

The worm commonly employed in the production of silk, is by Count Dondolo called the silk-worm of four moultings, of which he mentions two varieties: 1st.—That which makes a straw-colored cocoon, and that which produces a deep yellow one, and gives the preference for the former, the small silk-worm of three moultings. *The eggs of this species weigh one-eleventh less than the eggs of the common silk-worm— 39,168 of the latter weighing one ounce; while 42,260 of the former are required to make that weight. The silk is of a fine quality and takes 400 cocoons to make one pound, whilst 240 cocoons of the common silk-worm weighs that amount.

2nd.—The large silk-worm of four moultings, the eggs of which species the Count obtained from Friuli, is rather objected to on account of the coarseness of the silk.

3d.—The worms that produce white silk: In respect to this species the Count says: "I have raised a large quantity of them, and found them equal in all respects to the common silk-worm of four moultings, that producing white silk preferable to any other. If I raised silk-worms for my own spinning I would only cultivate the silk-worm of three moultings, and the white preferable to all others; and every year I would choose the whitest and finest cocoons, to prevent the degeneration of the species." This kind was introduced into France about the year 1783, and is there highly esteemed. It is supposed to be that which is known under the name of the "white worm," which produces two crops in one season.

---

* This means when green.

To the species enumerated by Dondolo it will be necessary to add:

4th.—The dark drab colored worm: This is very common in the United States.. They are commonly called the "black worm." They live longer and make a greater quantity of silk than the large white worm.

5th.—Silk worms of eight crops: There are two varieties, according to a statement of Lord Valencia; he found one at Jungpore, Bengal, which was supposed to be indigeneous, and is called *"Dacey;" and another variety he called †"China" or "Manrassa," which yields eight crops.

6th.—The mammoth white silk-worm, which is said to be a very superior species, makes a large cocoon and silk of a fine texture as well as very strong.

The distinction between one, two, three, &c. crop of eggs is not by some well understood. It means this: The eggs of the one crop can be hatched successfully from the eggs of the previous year, kept over winter to the following spring; but the two crop eggs may be hatched first from the eggs of the previous year, and next from the first hatching of the season.

The three crop eggs will hatch successfully from the same season's eggs, in so many repeated times.

The eggs of the one crop will not produce worms until the following season.

To the species enumerated above may be added others that are either yet wild and too rarely seen, which produce from other trees.

1st.—The Pennsylvanian silk-worm. This kind was found by the Rev. S. Tullin. He says he discovered the aurelia of a caterpillar, and on examination it was found not inferior to

---

*Travels in India, 1802, 1806; vol. i, p. 78; London 1809.

†The last may be the kind mentioned by Arthur Young, who says he obtained a silk-worm from China which he reared, and in twenty-five days he had cocoons; and by the twenty-ninth day he had a new progeny feeding in his trays. He remarks that "they would be a mine of wealth to those who would cultivate them."—Annals of Agriculture, vol. xxiii, p. 235.

the silkworm in the quality of its silk. The cocoons were three inches and a quarter in length, one inch in diameter, nearly resembling a dried bladder, of a reddish-brown color; weight twenty-one grains; it was covered with floss. Though perforated by the moth it unwound in hot water by which it was tested.—*British Annual Register and the Silkculturist.* Mr. Chambers, of Uniontown, found it on an elder bush, the cocoon being as large as a goose egg. The editor of the *Silkculturist* says naturalists call it the "Attacus Cecropia of Linnæus," and it feeds on the current, elder, barbery, wild cherry and other trees. The foregoing is confirmed by G. B. Smith, Esq., in the *Silkculturist* for October, 1837. Mr. Smith was not very successful with those he tried. They would not feed kindly, and the moth flew away as soon as it escaped from the cocoon, and the silk could not be reeled.

2d.—The Virginian silk-worm. Mr. Sheppard, of New Haven, says the editor of the *Silkculturist*, p. 19, presented us with a speciman of the Bombyx Virginicanus, or the native 'silk-worm of Virginia. It was found in great numbers on the plantation of J. B. Gray, Esq., Stafford County, and is capable of enduring the most rigorous winter. The cocoons are found suspended to the red cedar, and yield a beautiful white silk of a strong thread.

3d.—The Tusser or Bughy silk-worm, a native of Bengal, exceeding in size the common silkworm. The silk is found in great abundance in Bengal and adjoining provinces. From time immemorial it has supplied a durable, though coarse silk, which is much worn by the Brahmins and other castes of Hindoostan. When full grown it is about four inches long, bulky, of a green color, lateral strips of yellow edged with red. When ready to spin it envelopes itself in two or three leaves of the jujube tree, the same on which they feed. These leaves form an exterior envelope, which serve as a basin to spin the cocoon in, which is then suspended by a thick silk cordfrom the branch of the tree. It remains nine months in the pupa, or chrysalis state, and three months in the eggs or caterpillar.

The moth measures from the extremities of its wings, five or six inches; the female eight inches. It immediately escapes. Mr. Latrelle says it is the same kind as the wild one of China.

4th.—The Arrindy silk-worm. This is the Bombyx Cynethia of the naturalist; it is peculiar to Bengal, and feeds on the castor oil leaf. The silk from them can only be spun, and when made up can be washed only in cold water, as hot water will destroy the fabric.

5th.—The Jarrao silk-worm. This variety is found in India; the cocoons are spun in the coldest months; the silk is of a dark color. The males, when hatched, fly away, and the females remain upon the asseen tree, (the terminula alata glabra Roxburgh.) They are not impregnated by the males bred with them, but in ten or twelve hours another flight of males arrive, and the females afterward deposit eggs on the branches.

6th.—The Emperor moth is deserving of attention on account of the beauty of its cocoons. It feeds on fruit trees or willows, and spins a cocoon in the form of a Florence flask. The silk is strong, closely woven, well gummed, and the appearance is of damask as to softness, and is as pliable as leather. The tortrix chlorana, gypsy moth, cream spot, tiger moth, dock weevil, puss moth, and many others, spin cocoons which cannot be reeled, but make a very fine and elastic silk when spun.

7th.—The Bombyx Chrysorrhœa spins in company with three or four hundred round the ends of two or three twigs and leaves, leaving space for them to retire and shut themselves in, covering it with layers of silk so strong as to stand against wind and rain.

8th, 9th.—Tsouenkien and Tyankien, silk-worms of China.*

---

*Du Halde mentions in the province of Chanboug a species of silk is found on trees in great quantities, which is spun and made into stuffs called kient-chou. The silk of the first is a reddish grey, the other darker, very durable, and washes like linen.—History of China, vol. iii, p. 359. Historie des Science les Acts des China par Maillatom 2, p. 434, note 38. Madam Lottius' Treatise on Silkworms, Paris 1757. Annals of Botany 2d, p. 104.

10th.—The Social Silk Nest Spinner of South America: Don Louis Nec saw growing on trees in Chrysaneingo, Tixtala, in South America, ovate nests of caterpillars, eight inches long, which the inhabitants make into stockings and handkerchiefs.

11th and 12th.—Chinese wild silk-worms of the fagara and ash tree, and the Chinese wild silk-worm of the oak.

The memoirs of M. P. d'Incarville speak of three kinds of the wild silk-worm; one feeding on the fagara or paper tree and ash tree, and another on the oak.

There are two kinds of ash trees in China—the tcheon-tchun and kiang-chun—the former the same as our ash.

These worms moult four times. The cocoons are said to be as large as an egg, and are not reeled but carded and spun. .

In the first volume of the American Philosophical Society of Philadelphia is a paper of the late Moses Barnham, in which is recorded experiments made with caterpillars.

PART FOURTH.

CHAPTER I.—DISEASES OF WORMS.

1st.—*Pattre* is known, 1st—by yellow tinge of worms; 2nd—lengthened spine, shape and wrinkled skin; 3rd—from its shape and stretched feet; 4th—it eats little and languishes.

CAUSE.—Is said to be excess of heat during the dormant state, and pressure of litter.

REMEDY.—Remove them to a healthy place; feed a due supply of tender leaves and preserve a uniform temperature.

2nd.—*Grassenie or fat.* This disease generally appears about the second moulting, rarely later—and is scarcely known in the fourth age. Symptoms: 1st—eat but do not digest the food; 2nd—swell and become bloated; 3rd—their bodies become opaque and of a greenish color; 4th—around the breathing aperture becomes of a citron or dirty white color; 5th—their skins tear from the least touch; 6th—they are covered with an oily humor; 7th—they appear disposed to obtain relief from distension and stretching their feet; 8th—acrid humor proceeding from it. The last stage of this is death to any worm it comes in contact with.

. · Causes.—Said by Mr. Nysten and Mr. Roberts to be too glutinous and substantial food, occasioning indigestion in the young worm. La Browse says it is neglecting to dry the leaves when wet by rain or dews, and a lack of fresh air.

Remedy.—Remove to a distinct place, give less quantities of nourishment, poorer leaves and keep a moderate temperature.

3rd.—*Luisette to Shine.* Few worms are attacked with this until after the fourth moulting. Symptoms are 1st—shining; 2nd—a clear red changing to a dirty white; 3rd—it observed, a vicious humor drops from its silk tubes; 4th—its body becomes transparent.

Causes.—By negligence of feeding. Symptoms proved by Mr. Nysten, reseherces in Malardus, des verf a soie par R. N. Nysten, Paris, Des Vers a Soie, par te Reynard, Paris, 1824.

Remedies—Count Dondolo and M. La Brouse suggest instant removal; supply food gradually; increase slowly until perfectly restored.

4th.— *Yellows.* About fifth age. Symptoms: 1st—when about to spin body swells; enlargement of rings; the feet have the appearance of being drawn up from the tumesience of the adjacent parts; 2nd—appear of a yellowish color; 3rd—ceases to eat and runs about leaving stains of a yellow fluid; 4th—the yellowness first appears around the spiracula, and is diffused over the surface; 5th—becomes soft and bursts; 6th—the humor issuing becomes fatal to all worms it touches.

CAUSES.—Sudden exposure to great heat, also to moist or damp weather.

REMEDIES.—Instant removal'as before or change of air; fires if necessary ; oak leaves have been given with success.

5th.—*Muscardine numbness*, appears in fifth age. Symptoms: 1st—black spots in different parts of body; 2nd—become yellow and finally red or crimson color all over the body. 3rd—becomes hard and dry.

CAUSES.—Continuance of hot, dry and close air.

REMEDY.—Remove and purify the air by fumigation, and promote active circulation by ventilation.

6th.—*The Tripes* was first discovered by M. Regand de Lisle, of Crest. Symptoms: When dead they become placid and soft, and preserve a fresh and healthy appearance.

CAUSE.—Rainy weather. M. Nysten's experimental proof says rapid exhalations from the litter of an unclean cocoonery.

REMEDIES.—Dry air; flash fires in stoves, keeping at proper temperature. The discovery and remedy of this disease is ascribed to Mr. G. B. Smith, of Baltimore. Chloride of lime as a fumigator put in an earthern pan or cup.

CHAPTER II.—COCOONERY.

A cocoonery may be made of any dimensions, according to what the culturist may require.

It is necessary, before entering into sericulture, that some place or building should be prepared for rearing and feeding the worms; also for the spinning and making of cocoons. To the novice, or those who wish to commence on a small scale, I would recommend the fitting up of a small room, or corner of a room, that is well ventilated, with shelving made as follows: A frame 4 x 2 feet, with laths nailed across at a distance apart of half an inch. The cocoonery may be built any dimensions, with either adobe, brick or lumber, according to the number

of worms to be fed.   It would be better not to build too large
a room, but as the business increases, erect adjoining places, as
it would be better to have two small rooms than one large one.
It is estimated that a room fifty-four feet long, twenty-two feet
wide and ten feet high, would be large enough to raise over
twelve ounces of eggs.

The way to fit up said cocoonery is to make three rows of
frames for shelving all along, the posts to be placed four feet
apart, each way, divided in the middle so as to make the
shelving four by two, and placed one above another, fourteen
inches apart, commencing at least one foot from the floor, and
at a suitable distance from the pitch.   The frame work should
be made of scantling, planed for uprights, two by four.   The
room should not be full of windows but well ventilated at the
sides, near the bottom, and a ventilator on the top; it should
also be fitted up with a stove in a case.   After the hatching
out of the young worms the temperature of the room should
not be allowed to go down below 65°, or it will chill the worms
and thereby cause sickness or disease.   Let it, if convenient,
be built upon rising ground, removed from stagnant pools of
water, or any stench, and care should be taken that no refuse
is allowed to remain in the cocoonery, that tobacco smoke or
any other offensive matter is kept away, and that a thermom-
etre is kept to regulate the heat, which will be treated upon in
one of the succeeding chapters.

PART FIFTH.

CHAPTER I.—ON FEED FOR THE WORMS.

There are five different substances in the mulberry. 1st.—
the solid or fibrous; 2d—the coloring matter; 3d—water;
4th—saccharine; 5th—resinous.

The fibrous substance, the coloring matter and the water, excepting that which composes the body of the silk-worm, cannot be said to be nutritive to that insect. The saccharine matter is that which nourishes the insect, and that enlarges it and forms its animal substance. The resinous substance is that which, separating itself gradually from the leaf, and attracted by the animal organization, accumulates to itself and insensibly fills the two reservoirs and silk vessels which form the integral parts of the silk-worm. Therefore, from the different proportions of elements which compose the leaf, it follows that cases may occur in which a greater weight of leaves may yield less that is useful to the silk-worm. "Thus the leaf of the black mulberry, hard, harsh and tough, produces an abundance of silk, the threads of which are very strong, but coarse. The leaf of the white mulberry tree, planted in highlands, exposed to cold, dry winds, and in light soil, produces a large quantity of strong silk, of the purest and finest quality. The leaf of the same tree, planted in damp situations, in low grounds, or in stiff soil, produces less silk, and of a quality less pure and fine." The less nutritive substance the leaf contains, the more the worm must consume to complete its development; in consequence, the worm, from its fatigue, by taking less nutritive substance in a greater amount of leaves, would be more liable to disease than the worm which feeds on a more nutritive kind. The same may be said of those leaves which contian sufficient nutritive matter, but less of resinous substance; in this case the insects would thrive and grow, but probably would not produce either a thick or strong cocoon, proportionate to the weight of the worm.

State of Leaves for Feeding.—All silk growers have recommended the feeding with dry leaves free from both dew and rain. Nothing is so obnoxious to the insect as wet leaves. Count Dondolo, one of the greatest experts on the silk-worm, remarks: "These insects would be injured by eating leaves moist with either dew or rain."—*p.* 39. "The stripping of the leaves should not be begun before the disappearance of the dew, and ought to be concluded before the setting of the sun; it is all important to have always a supply of dry leaves."— *Count de Hazzi, p.* 65-67. "The preservation of the health of silk-worms depends essentially on the leaves being perfectly dry when given to them. Wet leaves invariably produce a diarrhea.—*Manuel published by order of Congress in* 1828, *p.* 128.

It it needless to multiply authorities on this subject. The worst leaf that can be given the silk-worm, and which has always injured it, is that which is termed "manner," and which arrives from the diseased state of the trees. The blighted or rust spotted leaves do not injure. The worm will eat this leaf, carefully avoiding the spots.

Preserving the Leaves.—To avoid these accidents, and to supply a resource for many days, a stock should always be kept on hand, sufficient for two or three days, which may be kept without injury in cold places, sheltered from the light, but not too dry, such as cellars, storehouses, back floors, &c. They would lose their dampness in too dry a place and should not be in one too damp. They should not be heaped up together too much, so as to promote fermentation.

Mode of Gathering Leaves..—Count Nerrie recommends the passing of the hand from the lower part of the branch to the top, and to strip the trees of its leaves upwards and not downwards, as the latter would injure the buds. This should be particularly enjoined on children and others who are employed in picking.

CHAPTER II.—HATCHING.

The hatching should not be attempted until the leaves of the mulberry are fully developed, so as to promote an abundance of foliage. It is always safer to be a few days late than a day too soon. The method in hatching of eggs pursued at Brooza, says Mr. Rhind, is: "The temperature of the chamber or cocoonery, or any other place used for the purpose, should be 63° to 64°; this is effected by the increasing of the fire or reducing by opening the windows, etc. This temperature should be carefully maintained for two consecutive days, and on the third increased to 66°; fourth to 68°; fifth to 70°; sixth to 72°; seventh to 75°; eighth to 77° and ninth to 81°. This goes to show that it is not mere precision that is here essential to success, but rather a gradual elevation of temperature to that maximum of heat which a transition from the egg to the larva requires. When the eggs are carefully exposed to heat in the manner described, they will show signs of vitality from the seventh to the ninth day. Count Dondolo, says: "The following are the signs of the speedy vivification of the silkworm. The ash-grey color of the egg grows bluish, then purplish; it then again grows grey with a cast of yellow, and finally of a tingey white. The young larva resembles a small black worm, and generally appears from sunrise to ten o'clock in the morning. It is important to keep each day's hatching by itself, by placing leaves over it. They may be easily separated from the eggs and put upon the hurdles or tables where they are meant to be raised, and kept apart, if one day's hatching, so as to be together through the season. Never should the worms of two consecutive days' hatching be put together, on account of their various stages of moulting. At times they will travel far, if kept with food. The silk-worm, if not properly attended to, is subject to disease. The congressional report on the silk-worm enumerates eight causes of disease in them, viz:

1.—Errors in hatching of eggs and in the treatment of very young worms.

2d.—Unwholesome air of the district in which they are bred.

3d.—Impurity of the air in which they are kept, arising from imperfect ventilation or from the exhalation of the litter and foces of the worms, which have been permitted to accumulate.

4th.—Too close crowding, owing to which cause their spiracles or breathing orifices are vented.

5th.—The quantity and quality of food.

6th.—Improper change of food.

7th.—Peculiar constitution of the air in certain seasons, against which no precaution can avail.

8th.—Frequent changes of temperature in the rooms in which they are kept.

Before proceeding any further it would be well to speak of the many enemies to which the silk-worm is subject. The hurdles or shelves or tables should not be put near the wall, and the walls of the room should be swept clean and kept free from cobwebs, as the spider is very destructive to the worm in its earliest stages, as are also ants, mud wasps, mice, birds, toads, poultry, and many other insects.

CHAPTER III.—FIRST STAGE AFTER HATCHING.

Rearing and Feeding.—Count Dondolo, in giving his precept as to the successful rearing of the silk-worm to the cocoon, says: "I must suppose that the silk-worms are kept until the first moulting at 75° of temperature, between 73° and 75° until the second moulting, between 71° and 73° until the third, and lastly, between 68° and 71° until the fourth moulting. One of the foundations of the art of rearing the silk-worm is to know the various degrees of heat in which the silk-worm should live; if this precept be not enforced, nothing can be performed with exactness." It is asserted in Mr. Roscoe's course of agriculture, that it is not relative to heat suitable to the condition of these industrious insects. It cannot be said that silk-worms are injured by any degree of heat in this climate, however considerable it may be. But a sudden change from moderate to violent heat, or the reverse, is injurious. On feeding Mr. Compstock says: "Though we have not much faith in arbitrary mathematical rules, yet as they may be of some probable use to the culturist in ascertaining the amount of food for his family of worms, we give them in such extracts from the manual published by authority of Congress in 1828. In doing this we will give the prescribed amount on each consecutive day of their life, with regard to the day of their respective ages:

It is stated by Dr. Lardner that it takes hurdles of eight square feet of space in the first age of the worms hatched from one ounce of eggs; and in the second age, fifteen square feet; third age, thirty-five square feet; fourth age, eighty-three square feet; and in the fifth or last age, one hundred and eighty-four square feet; the worms having four moultings. Having completed their growth in about thirty-three days the hurdles should be numbered 1, 2, 3, 4, etc., and each day's hatching kept separate, and fed as follows: First day, at intervals of two hours apart, three-quarters of a pound of chopped leaves cut very fine and sprinkled on the new hatched worms, which are one-twelfth of an inch long." "Give them plenty of room, feed them regularly four times a day," says Count Dondolo, "if you wish to have strong and healthy worms." Second day, give one pound and a quarter of chopped leaves; feed them at four different intervals, dividing the time equally, and giving the smallest quantity at the first feeding, and so increase gradually. Third day, give four meals, consisting of three and one-fourths pounds, cut fine as before named; they will now begin to turn a sort of hazel color, and have the bristly appearance of varnish. When viewed through a convex lens, their surface looks something like mother of pearl —transparent. Fourth day, as the worms approach moulting, a diminution of appetite occurs; let the first meal be about three-quarters of a pound and one pound and a quarter divided at the other meals; give them plenty of room whilst moulting, so as to avoid them sleeping in a crowded state, by gently separating and spreading them some. At the beginning of this day the first appearance of change is indicated, the worms begin to shake their heads and thus express uneasiness at the increasing tension of their skins, some scarcely eating any, keeping their heads in an elevated position; their bodies appear transparent; those nearer the moulting time, when seen against the light, are of a livid yellow tinge, but the greater number at the close of the day appear torpid and cease to eat. Fifth day, the young leaves chopped as before; about half a pound should be scattered thinly over them at four, and several times towards the end of the day; as a general thing the worms are torpid, and begin to revive. After this moulting they should be cleansed from their debris, so as to keep them healthy and strong after they have recovered. *Be sure to use the young leaves only in this stage.*

CHAPTER IV.—SECOND AGE.

About fifteen feet square of hurdles will be required for this numerous family during the second age until after their moulting. The temperature should be kept through this age at from 73° to 75°. The insects should not be lifted from their litter until they have all recovered; then lay a few leaves over them to crawl upon, and so distribute them on the shelves or hurdles. This will give them plenty of room to grow and they will not crowd each other. Sixth day—give now two pounds of young tender shoots and leaves, cut a little coarser than before, at intervals as before named, and after removing them to clean shelves or hurdles, thoroughly cleanse those from which they have been removed. Experience has proven that the silk-worm likes the tender boughs so much that they remain covered on them even when the leaves are consumed. Eighth day, give now, well chopped and picked, six and one-half pounds of leaves distributed at intervals as before, letting the two first meals be the largest, as some will begin to show symptoms of the second moulting, by the usual prognostics of rearing their heads and declining to eat. Ninth day, give one pound six ounces of picked, tender leaves, chopped small, and distributed as in the moulting before, lightly over them; on this day is again discovered the period, through their restlessness of change, and sinking under a stupor. The next day their old wardrobe is disposed of and they becomes as eager, or more so in their third life than in the first. Their color has now become of a light grey; the hair has become so much shorter as to be hardly perceptible to the eye. The muzzle, which in the first age was very black, hard and scaly, became immediately on moulting white and soft, now again becomes black, shining and scaly as before; and as the insect becomes older at each moulting, its muzzle hardens, because it needs to saw and bite larger and older leaves.

CHAPTER V.—THIRD AGE.

Tenth day give six pounds of tender shoots and six pounds of leaves, chopped small; at the close of this age they may be chopped more coarsely. The worms that have accomplished this age should not be removed from their shelves or hurdles until they are nearly aroused, for part will arouse on the ninth and part on the tenth day. No injurious consequence will occur if the part that has revived should wait twelve or fifteen hours until the rest are ready. A never-failing sign that they are all revived is the undulatory motion they display with their heads horizontally thrown over the shelves, so as not to foment, which would cause noxious evaporations and disease. The worms should now be removed as before, and will occupy thirty-five feet square of shelves or hurdles. Eleventh day—give at separate meals eighteen pounds of chopped leaves; the first meal should be the least— the worms will explain the reason themselves, as in the latter

part of the day they are voraciously hungry. Twelfth day —
nineteen and one-half pounds of picked leaves will be wanted,
chopped and divided; and the usual meals being given, towards evening their hunger begins to abate—therefore the last
meal should be the least. The worms will now grow fast, their
skins become whiter, their bodies semi-transparent, and their
heads longer; they will make various contortions as their
change approaches. Thirteenth day—ten and a half pounds
of leaves will be sufficient, chopped as before; give the usual
number of meals, the largest first and the smallest last, feeding
only those that require it. Should a great number be torpid,
whilst others require food, give only a slight meal without
waiting for the stated hours of feeding, in order to satisfy them
that they may sink into torpor speedily. Care of this kind is
important, and intermediate meals, given with discretion, will
prove beneficial. Fourteenth day—five and one half pounds
of leaves picked and chopped will be sufficient, in ordinary
cases more or less, as the case may require. Indications of
silk now begin to appear from the occasional depositions of the
insects. The worm now manifests inclinations to solitude and
free space to slumber in. It either climbs the edge of paper,
the elevated stalks of leaves, or failing in that, on the litter ;
it rears its head, expresses its uneasiness, and immediately on
the verge of the change it avoids all gross, excrementitious
matter, the only fluid remaining in the worms. This is that
which, prior to their change, gives them a yellowish-white
color, like amber. Whilst the worm thus prepares for the
moulting, sufficiently clear the air of the cocoonery by moderate ventilation. Fifteenth day—on this day the arousing of
the worms is an indication of the completion of the third age.
The muzzle of the worm during this age has maintained a
reddish color, it is no longer shining and black as it appeared
in the first age, but now becomes more lengthened and prom-
inent; the head and body are much enlarged since the casting
of the skin, or before they have eaten at all, a proof that they
were strengthened in the skin they have cast; and being now

unconfined, the natural specific density or rarefaction of their substance has expanded them at the ordinary rate of atmospheric pressure. At the completion of this age the body of the worm is more wrinkled, has become of a yellowish-white or fawn color, and without a glass no hairiness is visible in this age. A peculiar hissing noise is heard. This noise does not proceed from the action of the jaw, but the continual motion of the feet that sounds not unlike a shower of rain, until they fasten on the wood, when it ceases.

CHAPTER VI.—FOURTH AGE.

The worms with proper care, surviving from one ounce of eggs, require space equal to eighty-three square feet, and should be equally distributed, as already prescribed. The temperature should not be less than 68°, nor more than 71°, according to Count Dondolo; but when it rises, as at this season it inevitably will, higher compensating means must be sought by the instant removal of all litter liable to fermentation; give due circulation and ventilation throughout the cocoonery. After the third age be careful not to lift the worms from the hurdles until nearly all are aroused. It is, however, advisable to place the early roused in the coolest part of the building. The one part waiting a day or a day and a half for the other, as said before, is not injurious. Place the last roused in the warmest part of the building. Be it remembered that a moderate increase of heat sharpens the appetite and accelerates the growth of the worm, and vice versa. If the above rules are observed they will advance the maturity of the fourth age. Sixteenth day—give seven and a half pounds of the young shoots, and twelve pounds of picked leaves, coarsely chopped with a large blade. When the moment arrives for removing the worms from the hurdles, only one or two hurdles at a time should be covered with young shoots. These shoots, when loaded with worms, are afterwards put upon the empty shelves, and removed as in the first moulting. Many may remain upon the hurdles, who as yet have not strength to climb the young shoots, whilst those removed will have eaten all the leaves off the shoots; of necessity those remaining want to be fed a portion of the twelve pounds of leaves before named. At the end of this day the worms begin to evince renewed vigor. As they become more nimble, they lose their ugly color and become slightly white, and assume more animal vivacity. Seventeenth day—thirty-three pounds of leaves slightly cut up will now be wanted. The first meal should be the lightest, the last most copious. The worms now begin to grow fast, and their skins continue to whiten. Eighteenth day—forty-three pounds, slightly cut. The first meals of the day to be the most plentiful. Nineteenth day—fifty-one pounds, cut as before; the worms now begin to grow rapidly and reach one and a half inches long. Twentieth day—reduce to twenty-one pounds of the picked leaves, as the appetite diminishes; let the first meal be the largest and gradually lessen until the last; several are beginning to become torpid, with discrimination, only as they are wanted. The worms are now one and three-fourths inches long. Twenty-first day—of picked leaves give seven pounds, which are sufficient for this day. They now begin to decrease in size—they lose part of their substance before they sink into torpor. The greenish color of their rings becomes changed, and their skins are now wrinkled. Twenty-second day—the worms rouse on this day and accomplish their fourth age. In about the seventh day the worms have accomplished their fourth moulting. The insects are now assuming a darker color, or greyish with a red tinge. When the cocooneries are kept clean the air of the cocoonery is preferable to the external, from the odor of the fresh leaves.

CHAPTER VII.—FIFTH AGE.

This age of the silk-worm is the longest and most decisive. As they grow in this age they are liable to three evils: 1st—the quantity of fluid they disengage every day is occasioned by transpiration and evaporation of the leaves. 2d—the mephytic exhalations daily emitted from the excrementitious matter of the insects. 3d—The damp, as well as the hot state of the atmosphere of the cocoonery. The combination of these adverse circumstances may inflict injury upon the insects; the skin of the worm, by these means, is liable to relaxation and to lose its elasticity; they will also cause languor, decrease of appetite, and, unless the course be arrested, will cause sickness and death. The quantity of vital principle in the air is lessened by the increase of vegetable fermentation and fœtid exhalation, aggravated by the heat of the season. Therefore, it is very necessary to keep the cocoonery free from all vegetable and fœtid matter, so as to keep the worms healthy.

Twenty-third day.—At this time nearly all the worms are aroused, or have accomplished their fourth moulting. The cocoonery should be kept at a temperature of about 68 or 70 degrees. The worms should occupy about 102 square feet. Feed them eighteen pounds of the young shoots, or of common size leaves, not sorted, and also eighteen pounds of picked and sorted leaves. The eighteen pounds of shoots or other leaves are the ones on which the worms are to be removed; after they are all removed, the other eighteen pounds should be fed in four different meals.

Twenty-fourth day.—There will be required on this day fifty-four pounds of assorted leaves, divided into eight meals —the first to be the least and the last the most plentiful.

Twenty-fifth day.—The worms will now require eighty-four pounds of assorted leaves, divided into eight meals, the same as on the previous day.

Twenty-sixth day.—The worms will now require one hundred and eight pounds of leaves, fed as before. The voracious period of the worms is now rapidly advancing; some are two and one-half inches long.

Twenty-seventh day.—One hundred and sixty-two pounds of picked leaves will be wanted. It will be necessary for the worms to have intermediate feeds, when the regular distribution of leaves are devoured in less than an hour and a half. Otherwise, the worms need not receive any until the regular feeding time, which is every three hours.

Twenty-eighth day.— Give one hundred and ninety pounds of well assorted leaves—the first meal to be the largest —fed as before. Some of the worms will be now three inches long.

Twenty-ninth day.—Give one hundred and eighty pounds of well assorted leaves—first meal to be the largest—fed as before named, and diminished gradually. The extremities of the insects are now of a shining hue.

Thirtieth day.—The appetite diminishes this day, so that only one hundred and thirty-five pounds of leaves will be required, given in eight meals. The yellowish hue now extends from ring to ring, and they are gathering to the edges of the hurdles, which indicates their advancement to maturity.

Thirty-first day.—-Their wants now diminish and they require only ninety-nine pounds of leaves, to be distributed with care and discretion, as wanted.

Thirty-second day.—During this day the fifth age will be terminated, and the rising begins. Everything should now be cleansed and kept clean. They are now being perfected, which will be known by the following signs. 1st—instead of eating they get upon the leaves put upon the hurdles, raising up their heads ; 2d—when looking at them horizontally they appear of a whitish-yellow, transparent color; 3d—when they fasten to the inside of the edges of the hurdles; 4th—when they leave the centre of the hurdles, and crawl and try to reach the edges of the hurdles and crawl up them ; 5th—when their rings draw in and their greenish color changes to a deep golden hue; 6th—when their skins become wrinkled about their necks and their bodies feel like soft dough; 7th—when taken in the hand and looked through, the body of the worm assumes the appearance of a transparent, ripe yellow plum. These signs are prognostic of their rising. In preparing for them to spin, the usual plan is to put small oak or other boughs for them to crawl upon, and in places small pieces of paper, screwed up; but in the cocooneries of Europe they are fitted up with racks, as may be seen in Mr. Schettler's cocoonery in Salt Lake, and mine in Provo. The spinning is now fairly commenced.

CHAPTER VIII.—SIXTH AGE.

This age commences in the pupa state, and ends when the moth emerges from the cocoon.

The following are the necessary things that remain to be done. 1st—To gather cocoons. 2nd—To choose the cocoons which are to be preserved for their eggs or seed. 3rd—Preservation of cocoons until the appearance of the moth. 4th—The daily loss of weight which the cocoons suffer from the time they are finished until the appearance of the moths.

In gathering the cocoons care should be taken in removing them from the brush or racks, as the case may be, so as not to bruise or waste, and to save the floss. The floss should then be taken off with great care and delicacy, and the fibres not pierced, or the cocoons not flattened or bruised. Assorting the cocoons is to select those intended for seed that are perfect, whilst putting aside all imperfect or discolored. Fourteen ounces of cocoons are equal to one ounce of eggs, and one ounce of eggs will make one hundred and twenty pounds weight of cocoons. (That is to say green and not dry.)

An equal number of males and females should be selected. The male cocoon is smaller than the female one, depressed in the middle, as it were with a ligature, shaped in similar to a peanut, sharp in somewhat at both ends, with a great degree of hardness in those parts. The female cocoon is larger than the male, is round and full, not much and not often depressed in the middle, and more obtuse at both ends.

Preservation of Cocoons.—Where the temperature of the room is above 75°, the transition of the chrysalis to the moth state would be too rapid, and the coupling would not be productive. If below 68° the development would be too tardy Damp air will change it into a weakly and sickly moth. The apartment, therefore, should be kept in an even and dry temperature, between 66° and 73°.

The daily loss in weight of 1,000 ounces of cocoons, from the time of formation till the moth escapes: One day, 991, 992, 975, 970, 968, 960, 952, 943, 934, 925, according to the table prepared by Count Dondolo. In three or four days from the commencement of the spinning if the silk-worms have finished their cocoons, and in seven or eight days when they will be ready for picking from where they have been spun.* M. D' Homergue says eight days—but six days if there have been no thunderstorms to interrupt the labors of the moth. Dr. Parker informs us that with the use of electricity, his silk-worms have spun in twenty seven days from the hatchment.

The gathering should be performed with care, as much waste of silk is thereby saved. The cocoons may be gathered in five days from being finished, but where they do not all mount on the same day, it is possible that those may be culled that are not quite ripe with those that are. In gathering they should not be bruised, but carefully taken from their arches or twigs with all their floss.

Preservation of Cocoons Intended for Producing Eggs. —The temperature of the room should not be above 73° or the transition of the chrysalis to the moth state would be too

---

*Strip the cocoons clear of down or floss to prevent the feet of the moth becoming entangled when coming out.

rapid, and the coupling would not be productive. If below 66° the development is tardy, which is also injurious. Damp air will make them weak or sickly; therefore keep an even temperature, between 66° and 73°. Those not intended for raising the seed should be laid out for killing the chrysalides and drying.

Stifling the Chrysalides.—Where the quantity of cocoons is small, the necessity of cutting may be superseded by immediate reeling, the chrysalides should be destroyed between the fifth and twelfth day at furthest after the completion of the cocoon, or it will eat its way through and thus render the reeling of its work impracticable.

There are several ways of killing the pupa or chrysalis. 1st—by baking in an oven of the temperature of 88° or 89° wherein the cocoons are shut from four to six hours after being placed in bags, which must be turned occasionally, or moved to effect an equal exposure. 2d—by the sun's rays at a temperature of about 88° in which they may be left for three days from 9 o'clock a.m. to 4 p.m. 3d—by steam. For this purpose place the cocoons in a basket lined with three or four folds of woolen cloth to promote the equal dispersion of the steam; let the cocoons remain in a basket of sufficient dimensions to cover the mouth of the kettle, afterwards raise the basket on two pieces of wood, placed across the kettle, with water kept boiling over the fire. 4th—by suffocation in the gas from charcoal, which is effected by simply shutting the cocoons up for a night in a close room, wherein a pot of burning charcoal is placed. This last process is said to be the invention of G. B. Smith, of Baltimore, and to be the least injurious.

CHAPTER IX.—SEVENTH AGE.

It completes the entire life of the moth when the pupa, or chrysalis, has completed its transformation and is ready to depart, it puts forth a liquid, affirmed by some to be an acid, to dissolve the gum; the point being softened, it forces its beak through the fibres of the cocoon and makes its way into open day. At such times they should be spread upon a table or suitable place. They live after leaving the cocoon, from five to twelve days, according to the temperature to which they are exposed. The hours in which the moths burst the cocoons in greatest numbers are the three first hours after sunrise, if the temperature is from 64° to 66°. The male moth, the very moment he leaves the cocoon, goes eagerly after the female. When united, they should be taken up by the wings with great care, so as not to disturb them whilst coupled. When a shelf is fitted up with moths, the room should be made so that a person can hardly see in it. The hour of junction should be noted, and if any disunite they should be again brought together. Light injures them. The fluttering weakens and causes a loss of their vital and fecunding powers. The cocoons should be moved away as soon as left, and when the moth separates, after a sufficient time for conjunction, the males should be thrown away. During this period minute and constant attention should be paid.

The Separation and Laying of Eggs:

The male and female moths, at the proper time, if an opportunity is presented, will unite of their own accord; but when they do not they should be brought into juxtaposition, and after being coupled they should remain until they separate themselves. Some recommend separating them after six hours, but this is not consonant with the instincts of the insects, which are always the safest guide. If they separate prematurely they might again be brought together. Let the place be dark. The most vigorous of the males must now be placed with the unmated females. The females are not injured by waiting for the males a few hours—the only loss is a few unimpregnated eggs. While they are thus united, have ready clean white calico or sheets of white printing paper.

M. Deslongchamps, says he has used the male moth with success for six couplings, and the male was as lively and brisk as at the first disunion, which always had to be accomplished by the hand. The males, after leaving the females, if not wanted, should be thrown away, and after cleansing off the female she should be put on the cloth or paper to deposit her eggs, and there remain for from thirty to forty hours. If the cloth or paper is not covered, other females should be placed on the vacant spaces. The temperature should be from 65 to 80 degrees.

Preservation of the Eggs.

When the eggs have been deposited on the cloth or paper and have passed through the several changes of color, namely: after they have attained a pale clay hue, the cloth or paper must be folded so as to admit air into them, to prevent heating. The air should be dry and not above fifty degrees, and not below zero. Frost, it has been proven, does not hurt the eggs, but sudden changes do. They must be preserved from all other insects. vermin, and other enemies.

"Good keeping will produce good worms," said Mr. Rilley. If properly treated they will never degenerate in a dry climate.

NOTES.

To prepare Perforated or other Cocoons which cannot be reeled:

Fill a boiler with soft water; cut soap into shreds and dissolve in the water so as to make a strong suds; then put the cocoons into a bag made of mosquito bar, open at the side, so as to be able to take them out handily; let them remain in the water and keep it up to a scalding heat for an hour; then take them out and rinse in two or three clear waters; repeat until the gum is all removed; then dry and spin them on a small flax wheel, holding the cocoon between the finger and thumb; be sure to spin them even.

Last year, on account of late frosts, the early foliage of the mulberry trees got killed in some places. The consequence was that many lost their worms. Therefore, in future it would be well to sow in the fall the white cabbage lettuce, so as to be prepared with feed in case of such another occurrence, for the worms will do very well on the lettuce, during the first two moultings, but if you feed them with the mulberry leaf first they will not eat the lettuce.

In 1878 a great many worms were lost when they were ready to spin on account of the great heat, it being up to 80°. We lost nearly 60,000 when they had commenced to spin, they being in an upper room, and it could not be cooled by wetting the floor. Mrs. Wignell, of Payson, saved hers by making an experiment, as follows: She sprinkled on one of her tables over the worms cold water which caused them to revive, and served the others also the same way; she thus saved her worms. In speaking to Mrs. White, of Mill Creek, of the foregoing circumstances, she told me that was the way she did, and with the same result.